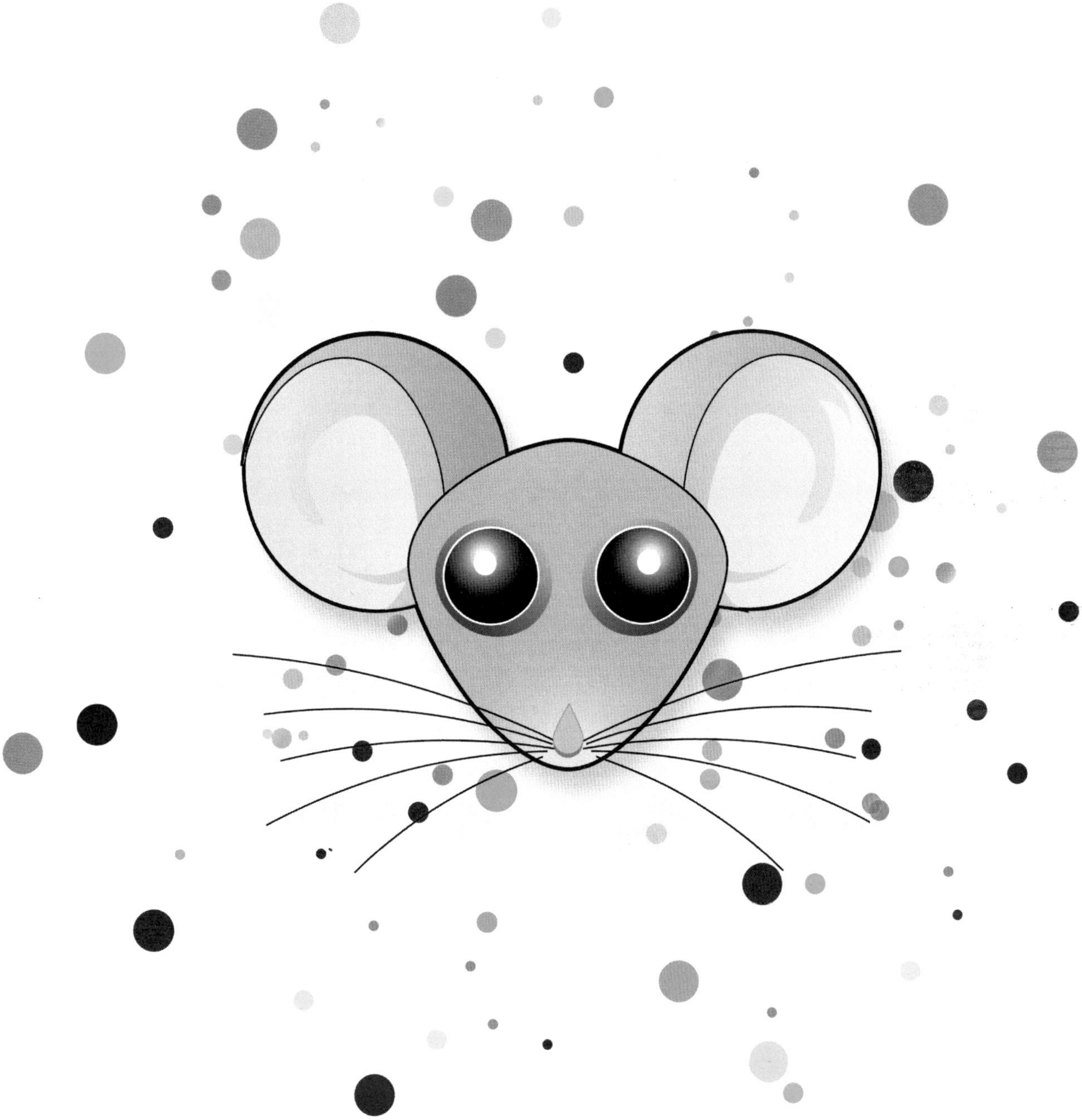

The author would like to thank those family members, friends and colleagues who have given their encouragement and support for this project.

A special thanks goes to Roberto Gonzalez roberto@rogolart.com | www.rogolart.ca for his wonderful creativity and expertise in illustrating and formatting this book.

9X FUN

For information about permission to reproduce selections from this book, contact Sharon Clark at sharon.clark@me.com

Printed in the USA

ISBN: 978-1495348594

By Sharon Clark
Illustrated by
Roberto Gonzalez

Multiply number 9 with me
It's really pretty cool to see

How simple finding the answers will be
When you discover this simple key

$9 \times 1 = 9$

$9 \times 2 = 18$

$9 \times 3 = 27$

Do you see a pattern here?
Inspect the answers 'till it's clear.

Add up the answers' separate bits.
Take good care and keep your wits.

0 + 9 = 9

1 + 8 = 9

2 + 7 = 9

"Aha", you say "I have a clue".
Let's try it with another few.

9 X 4 = 36
9 X 5 = 45
9 X 6 = 54

Add the answers' parts again.
The effort will not be in vain.

3 + 6 = 9

4 + 5 = 9

5 + 4 = 9

It seems that you are doing fine!
Each answer's parts add up to 9!
Let's do the rest. I think you're right.
How sharp you are - and oh so bright!

9 X 7 = 63
9 X 8 = 72
9 X 9 = 81

The rest do follow this simple rule.
Don't you think that's pretty cool?

6 + 3 = 9

7 + 2 = 9

8 + 1 = 9

1
5
9
2
8
7
4
3
6

Now one more time so that it's clear.
Watch the answers as they appear.

9 X 1 = 9	0 + 9 = 9
9 X 2 = 18	1 + 8 = 9
9 X 3 = 27	2 + 7 = 9
9 X 4 = 36	3 + 6 = 9
9 X 5 = 45	4 + 5 = 9
9 X 6 = 54	5 + 4 = 9
9 X 7 = 63	6 + 3 = 9
9 X 8 = 72	7 + 2 = 9
9 X 9 = 81	8 + 1 = 9

FUN

Wow! 9X math is really fun
Just one more thing and then we're done.

To get each answer really quick
There's one more very simple trick

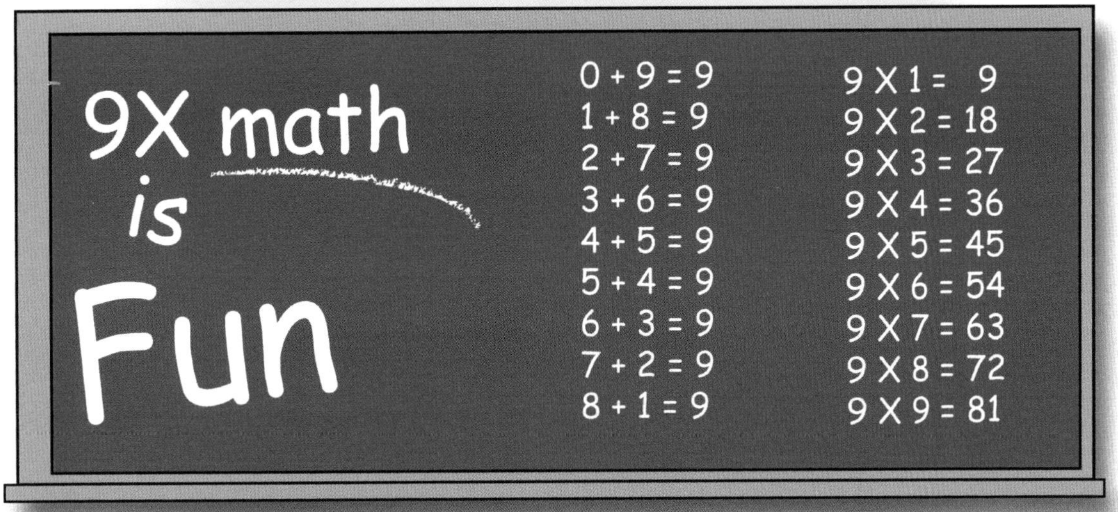

Can you spot another trend?
Check the equal signs at either end

Find each single number there
To see how both of them compare

9 X 1 = 09

9 X 2 = 18

9 X 3 = 27

9 X 4 = 36

9 X 5 = 45

9 X 6 = 54

9 X 7 = 63

9 X 8 = 72

9 X 9 = 81

9 X 1 = 9
9 X 2 = 18
9 X 3 = 27
9 X 4 = 36
9 X 5 = 45
9 X 6 = 54
9 X 7 = 63
9 X 8 =
9 X 9 =

You found the trend, but did you guess
Each number on the right is less?

By only one, I must confide
Than each number on the other side

So now we know another trick
To get the answer really quick

If someone asks, "What's 9 X 6?"
Just reach into your bag of tricks

6 - 1 = 5

6 less 1 makes 5, no doubt
You know what this is all about

Your answer starts with 5, that's fine
And has a number that adds to nine

That number must be 4 of course,
You know that from this fine resource

18 9
×
45 27
54 36
+
72 63
= 81

See how easy 9X is?
Your friends will think you're such a whiz!

Now you have some handy tools
For 9X math - just follow the rules

So go and have some 9X fun
That was easy. Now we're done!

9x

32416916R00015

Made in the USA
Charleston, SC
17 August 2014